THE POETRY OF CALCIUM

The Poetry of Calcium

Walter the Educator™

SKB

Silent King Books a WhichHead Imprint

Copyright © 2023 by Walter the Educator™

All rights reserved. No part of this book may be reproduced in any manner whatsoever without written permission except in the case of brief quotations embodied in critical articles and reviews.

First Printing, 2023

Disclaimer
This book is a literary work; poems are not about specific persons, locations, situations, and/or circumstances unless mentioned in a historical context. This book is for entertainment and informational purposes only. The author and publisher offer this information without warranties expressed or implied. No matter the grounds, neither the author nor the publisher will be accountable for any losses, injuries, or other damages caused by the reader's use of this book. The use of this book acknowledges an understanding and acceptance of this disclaimer.

"Earning a degree in chemistry changed my life!"
- Walter the Educator

dedicated to all the chemistry lovers, like myself, across the world

CONTENTS

Dedication v

Why I Created This Book? 1

One - Splendor Alone 2

Two - Cosmic Band 4

Three - Magnificent Strife 6

Four - Life Is Born 8

Five - Noble And True 10

Six - Steadfast And Strong 12

Seven - Essence Anew 14

Eight - Wondrous Substance 15

Nine - Calcium's Might 17

Ten - Eternal Light 18

Eleven - Creation's Grand Design 20

Twelve - Center Stage 22

Thirteen - Speaks Volumes 24

Fourteen - Oh, Calcium 26

Fifteen - Heart And Bone 28

Sixteen - Upreme And Free 30

Seventeen - Timeless Element 31

Eighteen - Crest Of Dawn 33

Nineteen - Defends 35

Twenty - Throughout All Time 37

Twenty-One - Marvel To Celebrate 39

Twenty-Two - Etches Its Name 41

Twenty-Three - Depths Of The Sea 43

Twenty-Four - Weathering Every Storm . . . 45

Twenty-Five - Everlasting Allure 47

Twenty-Six - Myriad Ways 49

Twenty-Seven - Tranquil Dream 51

Twenty-Eight - Testament To Fortitude . . . 53

Twenty-Nine - Rise And Shine 55

Thirty - Calcium's Story 57

Thirty-One - Let Us Cherish 59

Thirty-Two - Shimmer 61

Thirty-Three - Joy And Pain	63
Thirty-Four - Timeless Treasure	65
Thirty-Five - Teeth And Shells	67
Thirty-Six - Element Divine	69
Thirty-Seven - Mighty Calcium	71
About The Author	73

WHY I CREATED THIS BOOK?

Creating a poetry book about the chemical element of Calcium can be a unique and creative way to explore the properties, significance, and role of Calcium in our lives. Through poetry, you can convey the scientific aspects of Calcium in an artistic and engaging manner, making it more accessible and interesting to a wider audience. Additionally, poetry has the power to evoke emotions and connections, allowing readers to appreciate Calcium in a new light. This interdisciplinary approach can also foster a deeper understanding of science and art.

ONE

SPLENDOR ALONE

In the earth's embrace, you dwell unseen,
A silent guardian, steadfast and serene.
Calcium, oh noble element of might,
In bones and teeth, you weave strength and light.

From ancient oceans, you rose to the land,
A foundation for life, divinely planned.
In every heartbeat, you play your part,
Guiding muscles with a gentle art.

But beyond the science, a story untold,
Of a mineral so precious, a treasure to behold.
In the dance of stars, you found your birth,
A celestial journey, beyond the earth.

In the whispered secrets of the night,
You shimmer with grace, a celestial light.
A cosmic dancer, in the grand design,
A symphony of atoms, pure and divine.

Oh Calcium, in your atomic core,
Lies a tale of wonder, forevermore.
In the tapestry of existence, you shine so bright,
A celestial jewel, in the fabric of night.
So let us honor this element true,
For the wonders it holds, both old and new.
In every atom, in every bone,
Calcium, you reign, in splendor alone.

TWO

COSMIC BAND

In the silent dance of atoms, a celestial jewel reigns,
Calcium, the guardian of strength and light, it sustains.
In bones and teeth, it weaves its might,
A foundation of life, a celestial flight.

From ancient oceans to the land it came,
A symphony of atoms, a celestial flame.
In the cosmic dance, it takes its part,
A celestial dancer, a work of art.

With elegance and grace, it leads the way,
In the cosmic ballet, where stars hold sway.
A celestial journey, through time and space,
Calcium's splendor, in every place.

A silent guardian, a celestial guide,
In the fabric of life, it does reside.

A symphony of atoms, in every cell,
Calcium's presence, weaves its spell.
   So let us honor this element grand,
A celestial jewel, in the cosmic band.
In bones and teeth, it holds its light,
Calcium, a reigning element of splendor, shining bright.

THREE

MAGNIFICENT STRIFE

In the cosmic tapestry, a celestial dancer glows,
Calcium, a shimmering gem, in the heart of creation it flows.
Guardian of strength, bearer of light,
In the dance of atoms, it shines so bright.

From ancient oceans, it rose to the land,
A foundation of life, elegant and grand.
In bones and teeth, it weaves its grace,
A sculptor of form, a guardian in place.

In the symphony of elements, it takes its part,
Guiding the rhythm of life, with a delicate art.
Elegance personified, in the fabric of existence,
Calcium reigns, in splendid persistence.

A work of art, in the body's embrace,
A regal element, with poise and grace.

Honor its presence, in the grand design,
A reigning splendor, in the cosmic shrine.
    So let us marvel at Calcium's gleaming light,
As it dances through time, in the day and the night.
A celestial jewel, in the tapestry of life,
Calcium shines on, in magnificent strife.

FOUR

LIFE IS BORN

In the cosmic tapestry of existence, Calcium shines,
A shimmering gem in nature's grand design,
It sculpts the bones with an artist's grace,
Building the framework of life's embrace.

Silent guardian of strength and might,
In every cell, it weaves its light,
A regal element, noble and true,
Calcium, we owe so much to you.

In starry dance, you were born,
From the heart of a supernova torn,
A celestial dancer, graceful and free,
A cornerstone of life's symphony.

In every heartbeat, you play a part,
A vital rhythm, a work of art,
From the depths of earth to the heights of sky,
Calcium, your presence will never die.

So here's to you, element of grace,
In every atom, you find your place,
A shimmering gem, a sculptor of form,
Calcium, in you, life is born.

FIVE

NOBLE AND TRUE

In the cosmic dance of elements, you shine,
A regal gem in the celestial design.
Calcium, guardian of strength and light,
Weaving bones with grace, a wondrous sight.
    From the stars you came, a celestial birth,
Infusing life with your enduring worth.
In every cell, you play your vital part,
A symphony of existence, a work of art.
    Sculpting the framework of life's grand scheme,
You stand as a pillar, a celestial beam.
A silent force, yet mighty in your reign,
A shimmering gem that never wanes.
    In the tapestry of existence, you gleam,
A cornerstone of life's enduring dream.

So here's to you, Calcium, noble and true,
A radiant presence in all that we do.

SIX

STEADFAST AND STRONG

In the heart of every living frame,
Lies Calcium, a guardian, not tame.
Weaving bones with grace and might,
In the dance of life, a shimmering light.

A stalwart gem in the cosmic design,
Calcium, regal, pure and fine.
From stellar flames to earthly sod,
It stands as a symbol of strength and god.

A silent force in the web of life,
Calcium, a pillar in joy and strife.
In every cell, its worth holds true,
A radiant presence, old and new.

So here's to Calcium, steadfast and strong,
In the tapestry of existence, where it belongs.

A guardian of strength, a gem so bright,
Calcium, the essence of life's own light.

SEVEN

ESSENCE ANEW

In the cosmic dance, a regal star was born,
A guardian of strength and light, a gem adorned.
Calcium, noble and serene, in celestial design,
Weaving bones with grace and might, a presence divine.

In every cell, you stand, a silent force,
A symphony of existence, your enduring course.
Mighty Calcium, cornerstone of life's grand scheme,
Symbol of strength and god, in every gleam.

Oh, shimmering Calcium, essence of life's own light,
Enduring worth, steadfastness, in the cosmic flight.
A toast to you, O noble element, radiant and true,
In you, we find our strength, our essence anew.

EIGHT

WONDROUS SUBSTANCE

In the celestial dance of elements untold,
Calcium, a graceful dancer, shines bright and bold.
In the cosmic ballet, it takes center stage,
A guardian of strength, an eternal sage.
    Its elegance woven in the fabric of space,
A shimmering gem, a celestial embrace.
With steadfast persistence, it weaves its story,
In the grand tapestry of cosmic glory.
    From stellar nurseries to galaxies afar,
Calcium's presence, like a guiding star.
In the dance of creation, it plays its part,
A regal element, with beauty at heart.
    In the chambers of life, it finds its home,
In bones and shells, where its elegance is shown.

A vital essence, in the web of existence,
Calcium, a noble and wondrous substance.
    So let us marvel at its celestial grace,
In the cosmic ballet, it finds its place.
A guardian of strength, a shimmering light,
Calcium, a marvel, in the cosmic night.

NINE

CALCIUM'S MIGHT

In the cosmic dance of existence, behold
Calcium, a shimmering gem of old
Guardian of strength, in bones it weaves
Grace and might, in every sinew it cleaves
    A regal presence in life's grand scheme
A cornerstone of the body's gleam
In the tapestry of existence, it holds sway
As life's conductor in the symphony's play
    Radiant essence, in every cell it dwells
A noble element, the story it tells
Of resilience and endurance, it sings
In the orchestra of life, it spreads its wings
    Behold the beauty of Calcium's might
In the cosmic ballet, a radiant light
A guardian of strength, a pillar so true
In the grand design, it shimmers anew

TEN

ETERNAL LIGHT

In the whispers of time, Calcium reigns,
A regal element, in nature's chains.
A guardian of strength, with steadfast might,
Weaving bones with grace, in the veil of night.

In the cosmic dance, it shines so pure,
A symbol of strength, forever sure.
In the tapestry of existence, it weaves,
A silent force, in life's silent eaves.

Its presence divine, in the web of life,
A shimmering gem, amidst the strife.
A steadfastness, in the grand design,
An essence of light, so crystalline.

Oh, Calcium, in your regal glow,
You stand as a beacon, in the ebb and flow.
A guardian of strength, in the cosmic play,
A shimmering gem, in the light of day.

So let us honor, this element grand,
As it weaves its tale, across the land.
In the cosmic dance of existence, so bright,
Calcium, you shimmer, in eternal light.

ELEVEN

CREATION'S GRAND DESIGN

In the cosmic dance, a shimmering gem,
Calcium, the regal guardian, stands tall,
Symbol of strength, in the celestial realm,
A divine presence, reigning over all.

In the tapestry of existence, it weaves,
A radiant thread, with grace and might,
Binding bones with an enduring sheen,
In the symphony of life, a silent light.

In the alchemy of creation's grand design,
Calcium, the noble element, holds sway,
A testament to fortitude, pure and fine,
A beacon of resilience, come what may.

Oh, Calcium, in your luminous embrace,
We find the essence of power and might,

A celestial jewel, in the cosmic space,
A symbol of endurance, gleaming bright.
    So, let us honor this gem of the earth,
For in its presence, we find strength and worth,
Calcium, the regal guardian, divine,
A timeless emblem, in the cosmic shrine.

TWELVE

CENTER STAGE

In the cosmic dance of existence, Calcium reigns,
A regal glow in the symphony of life it sustains.
Noble and steadfast, it stands the test of time,
A testament to fortitude, a celestial sign.

Its enduring sheen, a beacon of resilience,
Binding bones with enduring strength, a silent brilliance.
In the tapestry of creation, it holds its place,
A symbol of power, endurance, and grace.

A celestial jewel, it sparkles with might,
A radiant thread, weaving through day and night.
In the grand design of nature, it plays its part,
A symbol of resilience, engraved in every heart.

So let us marvel at Calcium's enduring story,
A symbol of strength, in all its cosmic glory.

In the cosmic ballet, it takes center stage,
A testament to endurance, in every age.

THIRTEEN

SPEAKS VOLUMES

In the cosmic dance of existence, Calcium, noble and true,
A regal glow in the celestial queue,
Silent brilliance, steadfast and bright,
Guardian of strength, in day and night.

In the bones of the earth, your presence holds sway,
A foundation of might that will not decay,
In rivers and oceans, your essence flows free,
A mineral monarch, for all to see.

The whisper of your name evokes endurance untold,
A stalwart companion, in times hot and cold,
Through ages and eons, your story unfolds,
A cosmic architect, in mysteries untold.

In the grand design of life, you play a crucial part,
A beacon of resilience, from the very start,

Your quiet strength speaks volumes, in every living cell,
A testament to your power, in which we all dwell.
    Oh, Calcium, element of grace and might,
Your presence we cherish, in the day and the night,
In the fabric of existence, you stand tall and true,
Calcium, we salute the grandeur of you.

FOURTEEN

OH, CALCIUM

In the realm of elements, a mineral monarch reigns,
Calcium, the cosmic architect, in its silent strength sustains.
A beacon of resilience, it weaves through rivers' flow,
And in the depths of oceans, its tranquil might does show.

From ancient stones to living cells, its presence does endure,
Binding life and form together, steadfast and secure.
In bones and teeth, it builds its throne, a fortress strong and true,
A testament to its quiet power, in all that it can do.

Oh, Calcium, grand and noble, in your stately grace,
You stand as nature's cornerstone, in every time and place.

From the birth of stars to earthly realms, your essence does unfold,
A testament to your enduring might, a story to be told.
 So here's to you, Calcium, in all your grandeur bright,
A symbol of resilience, a force that shines so right.
In every particle that forms, in every living frame,
Your presence is a testament to nature's timeless claim.

FIFTEEN

HEART AND BONE

In the realm of nature's alchemy, a steadfast guardian stands tall,
Calcium, the silent architect of life's enduring thrall.
In bones and teeth, it weaves its might, a fortress strong and true,
A cornerstone of resilience, in every form and hue.

From ancient seas to mountain peaks, its story etched in stone,
A testament to timeless strength, in every living bone.
In whispered echoes of the past, it sings an ageless song,
Of power, grace, and steadfastness, enduring all along.

Within the tapestry of creation, it weaves its silent thread,
A symbol of endurance, where life and death are wed.

In every heartbeat, every breath, it dances through our veins,
A silent force of nature's grace, where power silently reigns.
 So let us raise a glass to Calcium, the guardian of our frame,
A whisper in the symphony of life, an unsung hero's name.
In every step, in every smile, its presence ever known,
A timeless ode to strength and grace, in every heart and bone.

SIXTEEN

UPREME AND FREE

In the cosmic dance of elements, you stand tall,
Calcium, mighty guardian, you never fall.
Within every bone, you weave a sturdy frame,
A fortress of strength, you eternally proclaim.

In the tapestry of existence, you gleam bright,
A beacon of light, in the darkest night.
Binding the fabric of life with your gentle embrace,
In every living cell, you leave a lasting trace.

O noble Calcium, symbol of power and grace,
Your endurance through time, none can efface.
With steadfast resolve, you hold nature's key,
In the grand design of life, you reign supreme and free.

From the depths of earth to the vast expanse of sky,
Your presence resonates, never to die.
In the symphony of creation, your story unfolds,
A timeless tale of might, as the universe beholds.

SEVENTEEN

TIMELESS ELEMENT

In the bones, you lie, O Calcium divine,
A fortress of strength, a guardian of time,
In every dance of life, you play your part,
Binding us together, a masterpiece of art.

    Mighty Calcium, in the earth you dwell,
A foundation of power, a story to tell,
From ancient seas to mountains high,
Your endurance echoes through the sky.

    In the whispers of the wind, your name is sung,
In the rivers and the streams, your legacy is flung,
You stand as a symbol of resilience untold,
A testament to nature's secrets, precious and bold.

    In every heartbeat, in every breath we take,
Your presence lingers, a silent, steadfast wake,

A silent force, a silent grace,
In your embrace, we find our place.
   O Calcium, in your quiet might,
You shape our world, a beacon of light,
So here's to you, in every form and hue,
A timeless element, strong and true.

EIGHTEEN

CREST OF DAWN

In the heart of bones, you find me,
Calcium, the sturdy cornerstone of life's decree.
I stand tall, unyielding in fortress and frame,
A silent guardian in the body's grand dame.

With steadfast grace, I weave the tapestry of strength,
Binding sinew to sinew, in a symphony of length.
I am the architect of resilience and might,
In every step taken, in every leap of flight.

From the depths of earth to the crest of dawn,
I endure, unchanging, since time was first drawn.
In the dance of atoms, I hold my noble ground,
An elemental force, in life's grand surround.

So sing of me, in praises strong and true,
For in every living cell, I'm a part of you.

A symbol of nature's power, steadfast and pure,
Calcium, enduring, forever to endure.

NINETEEN

DEFENDS

In the dance of atoms, you reign supreme,
Calcium, sturdy in the grand scheme.
In bones and teeth, you weave your might,
A lattice of strength, a scaffold of light.

But beyond the frame, your power extends,
In every living cell, your presence defends.
Enduring through time, a foundation of power,
In the body's grand design, you tower.

From the depths of the earth to the ocean's embrace,
You shimmer in nature, a silent guardian in place.
In coral reefs and marble halls, you hold your ground,
A symbol of resilience, in you, strength is found.

Calcium, oh Calcium, in your quiet grace,
You stand as a symbol of life's unyielding embrace.

In the grand design of existence, you play your part,
A mighty element, etched in every beating heart.

TWENTY

THROUGHOUT ALL TIME

In the dance of life, you stand tall and true,
Calcium, oh calcium, we sing to you.
In bones and teeth, you weave your might,
A guardian strong, a beacon of light.

In cells and nerves, you whisper your name,
Guiding the rhythm of the endless game.
Unyielding in strength, yet gentle in grace,
You're the foundation of every lively space.

In the tapestry of existence, you play your part,
Binding the threads with an unbreakable heart.
From the depths of the earth to the heights above,
You're a symbol of endurance, an emblem of love.

Oh calcium, oh calcium, we honor your role,
In every living being, you're a vital soul.

So here's to you, in this heartfelt rhyme,
A tribute to your power throughout all time.

TWENTY-ONE

MARVEL TO CELEBRATE

In the bones, you reside, oh mighty Calcium,
A fortress of strength, a guardian of form.
In every step we take, in every leap we make,
Your presence holds us steady, never to break.

A mineral so vital, yet often overlooked,
You stand as a symbol of resilience, never forsook.
In the dance of life, you play a vital role,
Building strong foundations, an essential soul.

In nature's grand design, you grace the earth,
In limestone cliffs and shells, you reveal your worth.
A testament to endurance, a symbol of grace,
In your steadfast presence, we find our place.

Oh Calcium, element of power and might,
You shape our world and make it bright.

In every heartbeat, in every breath we take,
Your strength and endurance, a marvel to celebrate.

TWENTY-TWO

ETCHES ITS NAME

In the garden of life, Calcium stands tall,
A silent guardian, strong and standing firm,
Binding the earth with its steadfast call,
In every cell and bone, its presence confirms.

From the depths of the soil to the ocean's sway,
Calcium weaves its magic, holding life in place,
A symphony of strength in its enduring display,
In every heartbeat, it leaves an indelible trace.

Like a sculptor shaping the body's frame,
It crafts the skeleton, a fortress so grand,
In teeth and bones, it etches its name,
A testament to resilience, in every land.

In nature's grand design, it plays its part,
Binding and holding, with a gentle embrace,

A symbol of endurance, from the very start,
Calcium, the cornerstone of this wondrous space.
    So here's to Calcium, steadfast and true,
In every atom, its power unfurls,
A silent force in all that we do,
A testament to strength in this magnificent world.

TWENTY-THREE

DEPTHS OF THE SEA

In the heart of Earth, deep and still,
Lies a mineral, steadfast, with an enduring will.
Calcium, ancient and strong, in rock and stone,
A silent guardian, steadfast and alone.

In the bones of giants, it weaves its might,
A foundation of strength, an everlasting light.
It whispers tales of resilience and grace,
In every step we take, in every embracing embrace.

From the cradle of life to the depths of the sea,
Calcium dances, unyielding and free.
A symbol of endurance, a beacon of power,
In every petal, in every budding flower.

In the symphony of life, it plays its part,
A silent conductor, with a gentle heart.

In every heartbeat, in every breath we take,
Calcium's presence, unchanging, awake.

So let us honor this element, noble and true,
For its quiet presence, in all that we do.
In the tapestry of existence, woven with care,
Calcium, eternal, beyond compare.

TWENTY-FOUR

WEATHERING EVERY STORM

In the quiet chambers of the earth, Calcium lies,
A silent guardian, hidden from prying eyes,
Its essence weaves through bone and sinew,
Granting strength and resilience anew.

In the dance of life, it plays a vital role,
Binding cells together, an unyielding soul,
In every heartbeat, in every breath we take,
Calcium stands firm, never to forsake.

In the hush of the night, under the moon's soft glow,
It whispers tales of endurance, in a language we know,
A symbol of grace, in its purest form,
Calcium, the steadfast, weathering every storm.

So here's to Calcium, the unsung hero,
In the tapestry of life, an eternal ground zero,

For in its embrace, we find love's tender touch,
A testament to its might, we cherish it so much.

TWENTY-FIVE

EVERLASTING ALLURE

In the bones, you reign, oh Calcium, guardian of form,
A silent sentinel, holding structure through life's storm.
In every step, in every leap, you stand strong and true,
A symbol of resilience, a testament to what we can renew.

You weave a lattice of strength, a fortress within,
Binding us together, protecting what lies within.
In the dance of atoms, you hold your ground,
A steadfast ally, in every cycle round.

Oh Calcium, in the grand design of existence,
You carve your mark with unwavering persistence.
From the earth's embrace to the beating heart's song,
You endure, you fortify, you belong.

So here's to you, oh mighty Calcium, steadfast and pure,
In the symphony of life, an everlasting allure.
A testament to endurance, a guardian so bold,
In your silent power, the grandest stories are told.

TWENTY-SIX

MYRIAD WAYS

In the symphony of life, you stand tall and strong,
Calcium, oh mighty guardian, to you we belong.
In bones and teeth, you weave a sturdy embrace,
A foundation of resilience, an enduring grace.

In cells and tissues, your presence whispers might,
A binding force, a silent guardian of the night.
In nature's embrace, you paint a vibrant hue,
A vital essence, a symbol of life anew.

Oh Calcium, you're the rock in the raging sea,
The steadfast anchor, the epitome of glee.
In every sunrise, in every moonlit night,
Your essence lingers, a beacon of pure light.

From the depths of earth to the heights of sky,
Your influence reigns, never questioning why.

In the tapestry of existence, you play a vital role,
A silent hero, an unwavering, steadfast soul.
    So here's to you, Calcium, in every form and phase,
For being the essence of strength, in myriad ways.
In every heartbeat, in every breath we take,
Your presence endures, an unyielding, resolute wake.

TWENTY-SEVEN

TRANQUIL DREAM

In the quiet realm of bone and shell,
Where strength and grace in silence dwell,
Lies Calcium, steadfast and true,
Binding life with a quiet, timeless hue.

In cells and tissues, it weaves its thread,
A silent force, where life is led,
Uniting, fortifying, standing tall,
In the grand design, it touches all.

As the world spins on its endless course,
Calcium whispers with unwavering force,
A symbol of endurance, a guardian's might,
In the dance of atoms, it holds the light.

In every heartbeat, in every breath,
Calcium lingers, defying death,

A sentinel of life, a fortress strong,
In its quiet presence, we all belong.
    So, let's raise a toast to this humble star,
In the grand tapestry, it's traveled far,
Calcium, the silent force of life's grand scheme,
A beacon of hope, a tranquil dream.

TWENTY-EIGHT

TESTAMENT TO FORTITUDE

In the quiet embrace of bone and earth,
Resides a stalwart guardian of life's worth.
Calcium, steadfast and unyielding in its might,
Weaves a tale of resilience in day and night.
    A silent sentinel, it stands tall and true,
Binding the fabric of existence, through and through.
In the dance of atoms, it takes center stage,
An unsung hero in life's intricate page.
    From infancy to age, it shapes our frame,
A silent architect, unknown to fame.
In every heartbeat, in every breath we take,
Calcium's presence, an unspoken ode to make.
    Amidst the chaos and the ceaseless flow of time,
It remains unmoved, a symbol so sublime.

In the grand design of existence, it plays its part,
A testament to fortitude, an enduring art.
    So here's to Calcium, in whispers and in song,
For being steadfast and resolute all along.
In the tapestry of life, its story shall unfold,
A mineral of strength, enduring and bold.

TWENTY-NINE

RISE AND SHINE

In the dance of atoms, you silently sway,
A steadfast force that never betrays.
Calcium, oh elemental king,
In the tapestry of life, your presence we sing.
    From bones to teeth, you fortify and bind,
Enduring and resilient, in every cell you're entwined.
A silent guardian, a structure so strong,
In the body's embrace, where you belong.
    In the quiet whispers of the night,
You hold us together, ever so tight.
A mineral marvel, so vital and pure,
In the dance of life, an unwavering allure.
    From the depths of the earth, you rise and shine,
A symbol of strength, in every design.

In the symphony of existence, you play your part,
A timeless companion, in every beating heart.
    Oh, Calcium, a marvel so grand,
In the fabric of creation, you firmly stand.
Endurance and resilience, in your silent decree,
A timeless essence, in the grand tapestry.

THIRTY

CALCIUM'S STORY

In the quiet core of life's design,
There lies a force both pure and fine.
Calcium, steadfast and unyielding might,
Weaves through existence, a guardian of light.

In every beat, in every breath,
Calcium dances, defying death.
Binding cells, a silent embrace,
In its endurance, life finds grace.

From sturdy bones to tender hearts,
Calcium plays its vital parts.
A warrior in the body's war,
It stands unbroken, forevermore.

In earth and stars, its presence known,
A testament to strength, not overthrown.
Silent force, yet power untold,
Calcium's story, forever bold.

Amidst the chaos, it holds the line,
A symbol of resilience, divine.
In the grand design, an anchor true,
Calcium, we owe our lives to you.

THIRTY-ONE

LET US CHERISH

In every cell, a silent force resides,
Guardian of light, where resilience abides.
Calcium, the binder of life's decree,
Enduring symbol of strength, it's plain to see.

From sturdy bones to tender hearts it weaves,
A lattice of fortitude, it never leaves.
In the earth and stars, its presence known,
A testament to endurance, brightly shown.

In the grand design of existence's art,
Calcium stands firm, playing its part.
A symbol of strength, unwavering and true,
An anchor for all, in all that we pursue.

So let us cherish this element so pure,
For in its presence, we can endure.

In every cell, in every living part,
Calcium's resilience, a beacon in our heart.

THIRTY-TWO

SHIMMER

In the heart of earth, you quietly reside,
A silent force, with strength inside.
Guardian of light, in bones you stay,
A symbol of resilience, come what may.
    In stars, you shimmer, a cosmic dance,
A spark of life, in the grand expanse.
Your presence endures, through ages untold,
A story of strength, in molecules bold.
    In the tapestry of life, you play your part,
Binding existence, with unyielding art.
A pillar of structure, in nature's embrace,
A testament to endurance, in every space.
    Oh Calcium, your essence profound,
In the symphony of elements, you resound.

A vital piece, in the grand design,
A symbol of fortitude, for all of time.

THIRTY-THREE

JOY AND PAIN

In the quiet of night, Calcium stands tall,
A guardian of light, it never falters or falls.
Beneath the earth's embrace, it silently thrives,
Binding life's decree, where resilience thrives.

    In bones and in teeth, it crafts its domain,
A silent force, in every joy and pain.
Enduring and strong, it holds life in its keep,
A symbol of strength, unwavering and deep.

    In the stars above, it traces its mark,
A shimmering light in the celestial dark.
From earth to the heavens, its presence is known,
A steadfast companion, in every life's zone.

    Through trials and triumphs, it remains ever true,
A binder of life, in all that we do.

So here's to Calcium, unwavering and bright,
A beacon of hope, in the darkest of night.

THIRTY-FOUR

TIMELESS TREASURE

In the grand design of existence, you stand,
Silent guardian, unwavering and grand.
Calcium, mighty in your quiet grace,
A symbol of strength, in every place.

Bones and teeth, you fortify and mold,
In the body's temple, your story's told.
Endurance and resilience, your hallmark true,
In the tapestry of life, we honor you.

From ancient seas to mountains high,
Your presence whispers, reaching the sky.
In shells and corals, you leave your trace,
A testament to time, in every space.

In gardens green, you nurture the earth,
From soil to stem, celebrating your worth.

In every heartbeat, a rhythm so pure,
You dance with life, steadfast and sure.
    Oh, Calcium, in your quiet might,
You weave through life, a guiding light.
In the symphony of elements, you play your part,
A timeless treasure, in every heart.

THIRTY-FIVE

TEETH AND SHELLS

In the silence of the earth, you dwell,
Calcium, steadfast guardian of life's story,
Binding bones, a fortress of strength,
In the tapestry of existence, you shine with glory.
    From the stars to the deep roots below,
Your essence weaves through time's embrace,
A silent force, nurturing the soil,
Embracing life with unwavering grace.
    In the dance of light, you hold the key,
To resilience and steadfast might,
In teeth and shells, your mark is seen,
A symbol of endurance, a beacon of light.
    Oh, Calcium, emblem of fortitude,
You stand as a testament to nature's design,

In every heartbeat, in every breath,
Your presence weaves through the grand design.
    Through trials and triumphs, you remain,
A silent force, a guardian of the earth,
In your timeless essence, we find strength,
Oh, Calcium, in you, resilience finds rebirth.

THIRTY-SIX

ELEMENT DIVINE

In the earth's embrace, you silently thrive,
A force of nature, keeping life alive.
Calcium, binder of bones and shell,
In your quiet strength, we find our dwell.

Guardian of light, in the ocean's deep,
Your presence lingers, a secret to keep.
A symbol of resilience, enduring and bold,
In your steadfast nature, stories untold.

Through time you weave, a tapestry strong,
Binding life's decree, where you belong.
In every heartbeat, in every breath,
Your essence whispers, conquering death.

A fortifier and molder of the living frame,
In your embrace, we find our claim to fame.
A beacon of hope, a guiding light,
Embracing life with grace, oh wondrous sight.

Calcium, oh noble element divine,
In your silent power, we entwine.
As you stand tall, in nature's grand design,
You are the essence of strength, oh, how you shine.

THIRTY-SEVEN

MIGHTY CALCIUM

In earth and stars, your presence known,
Mighty Calcium, on your throne.
Enduring force, unwavering might,
Symbol of strength in purest light.

Binding nature, firm and true,
In bones and teeth, we find you.
Enduring trials, triumphs too,
Calcium, we honor you.

Silent force, in nature's scheme,
Nurturing life, like a dream.
Fortifying all you touch,
In grand design, you matter much.

Oh Calcium, your story told,
In whispers ancient, still so bold.
A mineral of timeless grace,
In every moment, every place.

So here's to you, Calcium dear,
In every atom, crystal clear.
Enduring, binding, nurturing too,
Mighty Calcium, we honor you.

ABOUT THE AUTHOR

Walter the Educator is one of the pseudonyms for Walter Anderson. Formally educated in Chemistry, Business, and Education, he is an educator, an author, a diverse entrepreneur, and he is the son of a disabled war veteran. "Walter the Educator" shares his time between educating and creating. He holds interests and owns several creative projects that entertain, enlighten, enhance, and educate, hoping to inspire and motivate you.

Follow, find new works, and stay up to date with Walter the Educator™ at WaltertheEducator.com

www.ingramcontent.com/pod-product-compliance
Lightning Source LLC
LaVergne TN
LVHW051959060526
838201LV00059B/3727